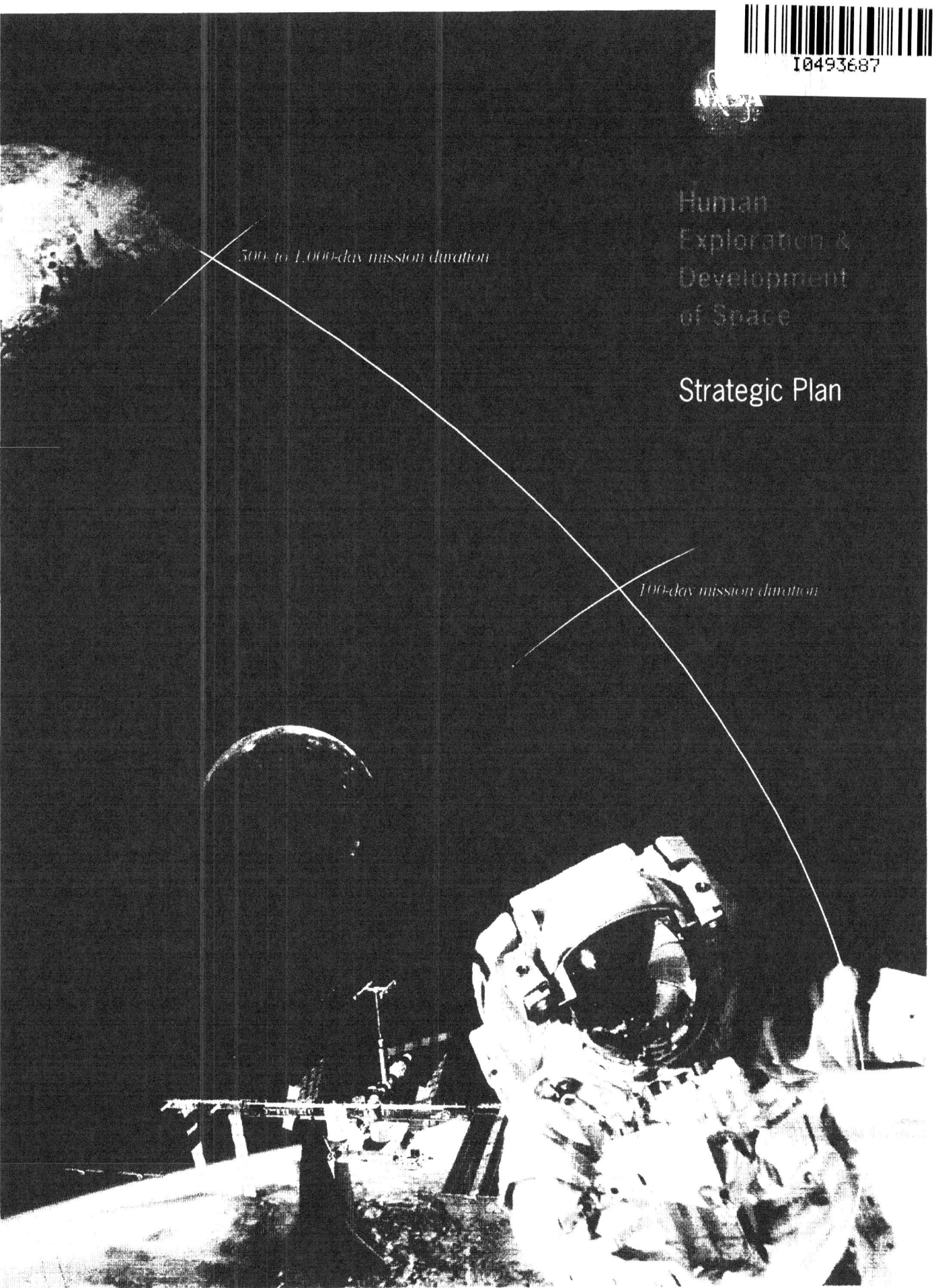

Human
Exploration &
Development
of Space

Strategic Plan

500- to 1,000-day mission duration

100-day mission duration

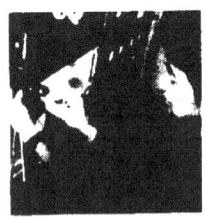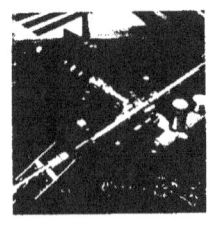

NASA is an investment in America's future.

As explorers, pioneers. and innovators. we boldly expand frontiers in air and space to inspire and serve America and to benefit the quality of life on Earth.

• To advance and communicate scientific knowledge and understanding of the Earth, the solar system, and the universe

• To advance human exploration, use, and development of space

• To research. develop, verify, and transfer advanced aeronautics, space, and related technologies

• To expand the frontiers of space and knowledge by exploring, using, and enabling the development of space for human enterprise

Letter From the Associate Administrators

Joseph H. Rothenberg

Dear Human Exploration and Development of Space team and stakeholders:

As we enter a new millennium, people the world over are reflecting on the accomplishments of the past and speculating about opportunities of the future.

Some of the most inspiring and important accomplishments of the past four decades have resulted from the space program: events such as the planet-wide impact of the Apollo landings on the Moon and images of Earth; discoveries such as the astonishing Hubble Space Telescope (HST) photos of solar system formation; achievements such as the sending of the first human artifacts—Pioneer and Voyager spacecraft—beyond our solar system; new capabilities such as communications and weather satellites; and advances in areas of fundamental and applied research ranging from basic physics to bone loss and aging to telemedicine. Space has touched the lives of many hundreds of millions worldwide.

The vision of our role in the new century is clear: NASA is an investment in America's future. We will boldly expand the frontiers in air and space to inspire and serve America and to benefit the quality of life on Earth.

During the coming 25 years, we must achieve profound strategic goals in space. In supporting the realization of that vision, the mission of HEDS is to expand the frontiers of space and knowledge by exploring, using, and enabling the development of space for human enterprise.

It is in this context that we present this new Strategic Plan for the HEDS Enterprise—a plan that we believe will enable us to achieve challenging technical feats with enormous societal benefits.

Our goals are fivefold:

- To explore the space frontier

- To expand scientific knowledge

- To enable humans to live and work permanently in space

- To enable the commercial development of space

- To share the experience and benefits of discovery

We begin with the foundation of the Space Shuttle and the International Space Station (ISS), now under construction in Earth orbit.

As we go forward, we will probe fundamental questions in science: What is the role of gravity in chemical and physical systems? What is the role that gravity plays in biological processes of both plants and animals? Can humans live and function productively in an environment away from the surface of Earth? We will enable humans to participate in probing questions, such as "does life exist elsewhere in the solar system?"

We also aspire to make possible U.S. leadership of international efforts to extend permanently human presence beyond the bounds of Earth, involving both machines and humans as partners in innovative approaches to exploration. We will engage the private sector in the commercial development of space in order to enable the continuation of current space business and the creation of new wealth and new jobs for the U.S. economy.

Together we can accomplish these goals and enable historic improvements in our understanding of nature, in human accomplishment, and in the quality of life. This Strategic Plan for the Human Exploration and Development of Space Enterprise is a first step.

Joseph H. Rothenberg
Associate Administrator,
Office of Space Flight

Arnauld E. Nicogossian, M.D.
Associate Administrator,
Office of Life and Microgravity Sciences
and Applications

Arnauld E. Nicogossian, M.D.

3

The Challenge

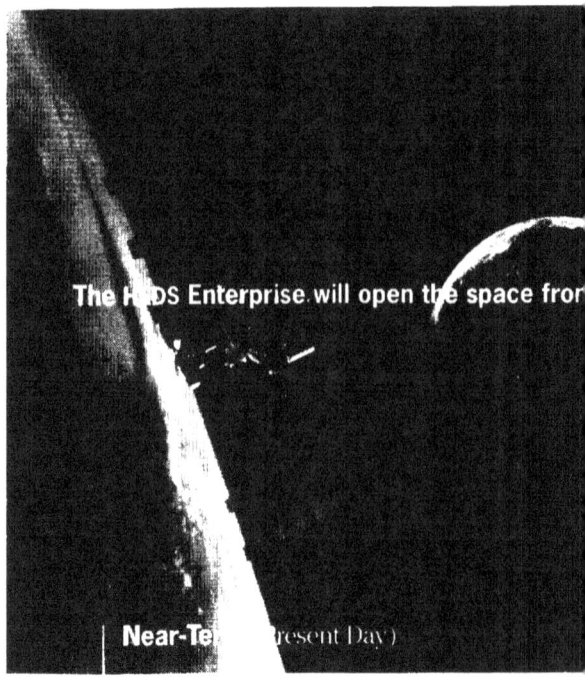

The HEDS Enterprise will open the space fron

Near-Te resent Day)

- Undertake, in partnership with the scientific community, sustained international explorations throughout the inner solar system by integrated human/robotic expeditions

- Achieve breakthrough discoveries and technology developments through basic, applied, and commercial research in the unique venue of space—exploiting characteristics such as microgravity, vacuum, radiation, and location

- Establish safe and routine access to space in support of permanent commercial human operations in low-Earth orbit and ongoing exploration activities at one or more sites beyond Earth orbit

- Engage the private sector in the commercial development of space and enable the creation of new space industries generating new wealth for the U.S. economy

- Communicate the excitement and importance of the discovery of new worlds and the profound insights we will gain into the origins of life and the universe

In order to guide planning, the HEDS Enterprise has identified some potential future targets and goals (e.g., "Design Reference Points") — beginning with the near-term and extending to the far-term and beyond.

Space Shuttle 7–14-Day Mission Duration, International Space Station 30–90-Day Missions

The HEDS Enterprise will continue to pursue privatization of Shuttle operations, and invest in upgrades and improvements to ensure that the Shuttle provides safe and reliable human access to space until a replacement system is available. The Shuttle and the ISS will be the foundations for continuing utilization and commercialization of Earth orbit. As the ISS becomes a reality, plans are under way to open the facility aggressively to commercial development. NASA is encouraging entrepreneurs and scientists to identify opportunities and to use the ISS for "Pathfinder" projects to bring business concepts to reality.

At the same time, the HEDS Enterprise will partner with industry and others to develop and demonstrate new technologies to enable future goals.

Design Reference Points

"100-day class" missions are those in near-Earth space, where the total mission duration is approximately 100 days; "1,000-day class" missions are those within the inner solar system, within 2 astronomical units (AU) of the Sun, where the total mission is approximately 1,000 days in duration. (One "AU" is defined as the Earth's average distance from the Sun—about 93 million miles.)

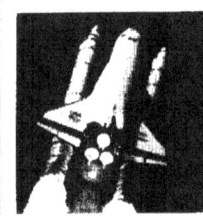 

Design Reference Points:
Low-Earth Orbit and International Space Station.

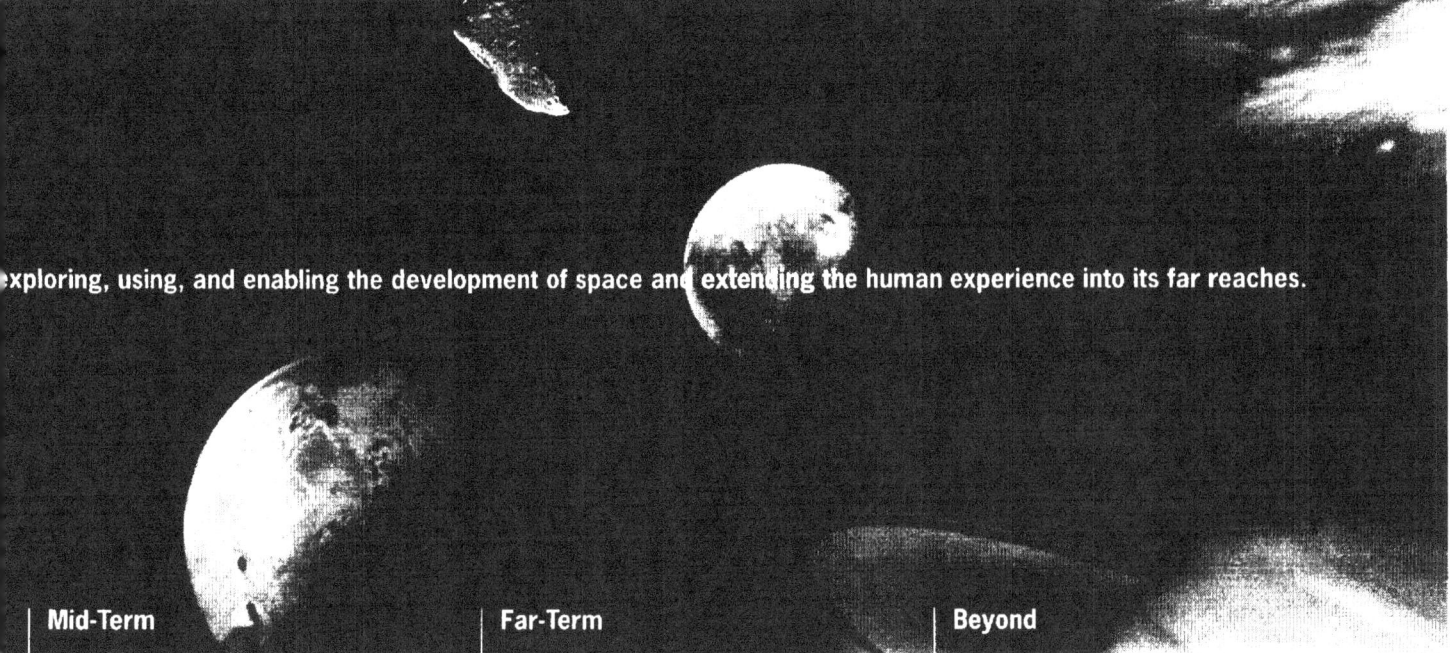

...xploring, using, and enabling the development of space and extending the human experience into its far reaches.

| Mid-Term | Far-Term | Beyond |

100-Day Mission Duration

In the 1960's, Apollo astronauts landed on the Moon and began the era of exploration beyond Earth orbit.

Missions to Earth-Moon and Earth-Sun "Libration Points" (where gravitational forces balance) could maintain revolutionary new telescopes and establish key infrastructures.

100-day mission capabilities would allow further exploration of the lunar surface to enable the expansion of knowledge and to provide an experience base from which we can reduce the cost and risk of further explorations. A lunar mission could also answer the questions of how we can use lunar resources commercially and how we can sustain operations in another planetary venue.

500- to 1,000-Day Mission Duration

Through space science mission payloads, HEDS mission designers will learn about the surface of Mars in preparation for integrated human/robotic missions early in this century.

Resolving fundamental questions about the history of Mars and the possibility of past or present life elsewhere in our solar system may be compelling reasons for such future missions.

In the region of space between Earth and Jupiter, there are many tens of thousands of asteroids—some composed of valuable minerals, others of materials that could be used to make propellants, in space construction or for commercial ventures. Advancing HEDS capabilities could make possible future human/robotic missions to these challenging and scientifically interesting targets.

2,000 Days and Longer Mission Duration

As technology advances, design reference points in the outer solar system—such as Ganymede, a moon of Jupiter, and Titan, a moon of Saturn that has an atmosphere similar to that of ancient Earth— might become accessible to human missions later in this century.

Although unlikely in the coming decades, eventually technology may open the way for major probes to the very edges of our solar system and beyond.

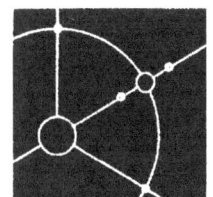 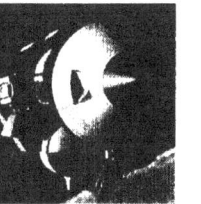

Design Reference Points:
Libration Points and Lunar System.

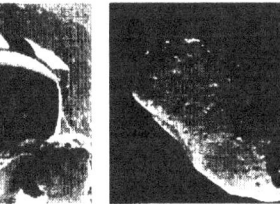 

Design Reference Points:
Mars System and Asteroids.

Design Reference Points:
The Moons of the Outer Planets and Beyond.

The Five HEDS Strategic Goals

- Explore the Space Frontier

- Expand Scientific Knowledge

- Enable Humans to Live and
  Work Permanently in Space

- Enable the Commercial
  Development of Space

- Share the Experience and
  Benefits of Discovery

This Strategic Plan explains
why and describes how.

An astronaut
participates in
Hubble Space
Telescope servicing.

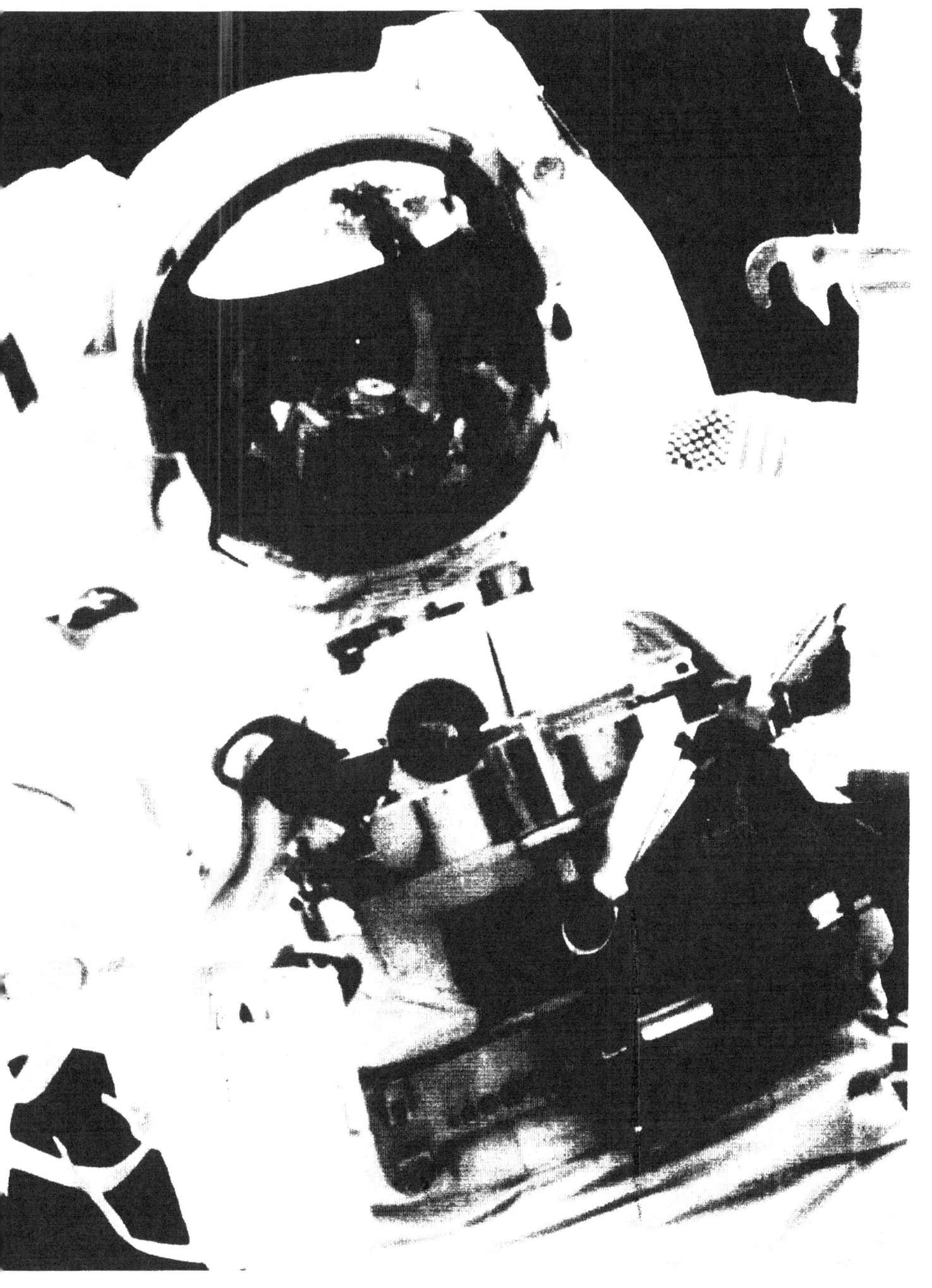

There are certain ideas that many believe to be inherent in the human psyche and integral to American culture: ambition for progress, curiosity about the unknown, the need to pose profound questions and to answer them, the concept of new frontiers that—once achieved—promise a better quality of life for all peoples. Space is such a frontier. Earth orbit, the Moon, near-Earth space, Mars and the asteroids, eventually the moons of the giant planets of the outer solar system, and someday more distant worlds—these are collectively the endless, ever-expanding frontier of the night sky under which the human species evolved and toward which the human spirit is inevitably drawn.

It is a fundamental goal of NASA to expand the space frontier progressively through human exploration, utilization of space for scientific research, and commercial development.

Our Strategic Objectives

- Invest in the development of high-leverage technologies to enable safe, effective, and affordable human/robotic exploration

- Conduct engineering and human health research on the International Space Station to enable exploration beyond Earth orbit

- Enable human exploration through collaborative robotic missions

- Define innovative human exploration mission approaches

- Develop exploration/commercial capabilities through private sector and international partnerships

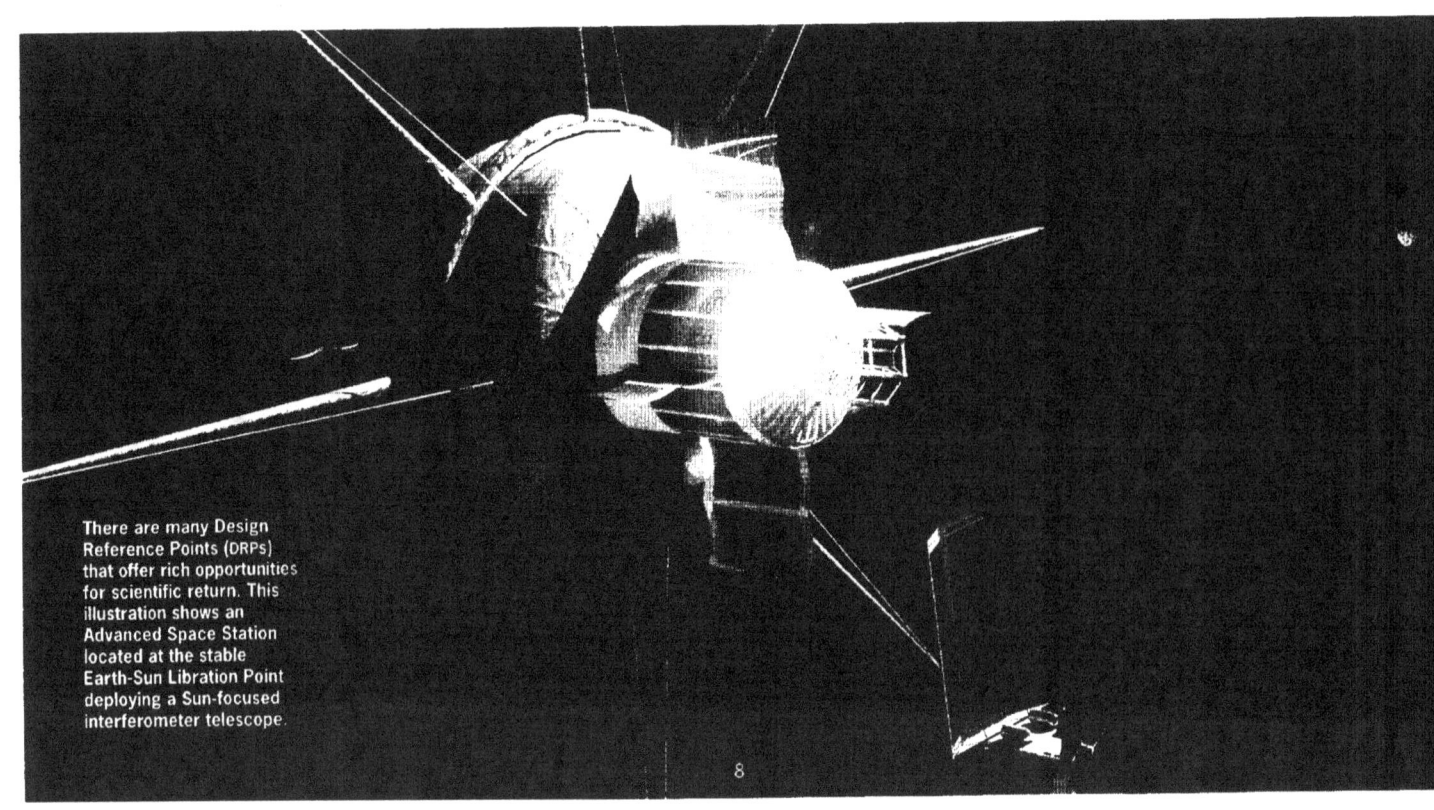

There are many Design Reference Points (DRPs) that offer rich opportunities for scientific return. This illustration shows an Advanced Space Station located at the stable Earth-Sun Libration Point deploying a Sun-focused interferometer telescope.

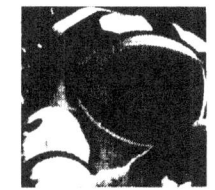

Time-Phased Objectives

• Probe environments beyond low-Earth orbit (LEO) in collaboration with the science community and obtain data needed for design decisions concerning future safe, effective, and affordable human exploration—this includes characterizing space and atmospheric environments; characterizing space "weather"; establishing geological characteristics; setting potential science goals; validating physical processes and technologies; and identifying planetary resources and potential landing sites.

• Obtain data in collaboration with the science community needed for design decisions concerning future safe, effective, and affordable human exploration—this includes establishing environments, space "weather" geological characteristics, and potential science goals; validating physical processes and technologies; and identifying planetary resources and potential landing sites.

• Initiate the development of the interdisciplinary knowledge base (biology, physics, materials, etc.) needed to enable safe, effective, and affordable human and robotic exploration.

• Test and validate countermeasures to physiological problems in long-term space flight.

• Identify and evaluate candidate approaches for 100- to 1,000-day human missions capable of a 5- to 10-fold cost reduction—while increasing safety and effectiveness (compared to 1990's projections).

• Identify with U.S. industry the commercial potential inherent in candidate 100- to 1,000-day human mission concepts and technologies.

• Develop and test—on the ground and in space—competing technologies for human missions beyond LEO in

cooperation with other agencies and international partners, and with U.S. industry (via jointly funded research and development (R&D) when mutual objectives permit).

• Collaboratively with the space science community, conduct ambitious robotic/engineering missions that establish continuing operations at key sites (e.g., "outposts") in order to:

- Acquire important engineering data and validate key technologies, while collecting critical data on the space environment (including "space weather")

- Build up and validate infrastructure needed for later safe and affordable human expeditions, and

- Enable highly-productive, unique science activities prior to the arrival of human explorers.

• Conduct applied research to improve understanding of and develop effective and efficient countermeasures for space radiation and the microgravity environment to enable safe and affordable 1,000-day human missions.

• Facilitate privately funded industry development of key 100-day human mission capabilities and identify (jointly with industry) the commercial potential of—and technology R&D projects for—1,000-day and longer human missions.

• Develop the capability for affordable, 100-day class integrated human-robotic expeditions to a previously established site beyond LEO (e.g., an "outpost") in collaboration with international partners to expand scientific opportunities dramatically and to test and evaluate advanced concepts and technologies for later capabilities and operations.

• Refine existing—and continue to identify and evaluate innovative new—approaches for 1,000-day missions capable of a 10- to 20-fold cost reduction

compared to earlier projections, and continue the development and validation of competing breakthrough technologies in cooperation with external partners, including U.S. industry.

• Complete the development of safe, self-sufficient, and self-sustaining systems that can enable humans to live and work in space and on other planets —independent from Earth-provided logistics—for extended periods.

• Pursue ambitious collaborative robotic/engineering missions that expand activities at existing and additional key sites (e.g., "outposts") beyond LEO.

• Develop the capability for the first 1,000-day, highly integrated human-robotic expedition to a previously established site beyond LEO (e.g., an "outpost") in collaboration with international partners to expand human knowledge and test technologies for the continuing extension of human activities and enterprise in space.

• Finalize candidate architectures and begin technology R&D to enable a further two- to fourfold reduction in costs for ambitious long-term human exploration objectives, making use of revolutionary technologies and both new and existing infrastructures.

Expand Scientific Knowledge

Throughout the history of scientific inquiry, gravity has been an inescapable and confounding influence. Every living organism has evolved under gravity's constant pull; every physical process we have studied, we have studied in the presence of gravity. Access to the environments of space and of the planets has led us to raise new and profound questions affecting our fundamental theoretical understanding of nature, including living systems. As we have gained control over crucial variables such as gravitational acceleration, a new window has opened on the world around us.

It is the goal of HEDS to take advantage of the opportunities afforded by space to expand our fundamental knowledge of physical and biological processes—first using relatively simple model systems and then increasingly complex ones. This approach will enable rigorous contributions to scientific knowledge in general, as well as longer term advances in technology of benefit to both NASA and commercial applications.

Our Strategic Objectives

- Investigate chemical, biological, and physical processes in the space environment, in partnership with the scientific community

- Expand collaborative research on the International Space Station that will further human exploration of the solar system

- Significantly extend scientific discovery on missions of exploration through the integrated use of human and machine capabilities

Scientific experiments in space today are expanding our knowledge to enable humans to live and work permanently in space tomorrow.

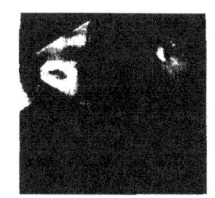

Time-Phased Objectives

- Investigate the gravity-related mechanisms affecting cellular biotechnology to enhance our understanding of Earth-based limitations and space-based possibilities.

- Use advanced optical techniques to probe structural development during formation or growth of organic and inorganic materials possessing enhanced properties.

- Use the space environment to examine combustion processes relevant to technology development for Earth and space-based fire safety, as well as for fuel efficiency, pollution reduction, and the combustion-based synthesis of novel materials.

- Lay the theoretical and experimental foundations for atomic physics investigations in the areas of laser cooling of atoms, atomic clocks, and biological physics.

- Begin long-duration studies on the ISS to understand the molecular mechanisms of cellular response to gravity and other characteristics of space flight.

- Initiate studies on the adaptability of Earth life to the space environment and on the signatures of life on other planets in our solar system.

- Discover fundamental principles underlying biological responses to space flight, using ground-based and flight research.

- Identify important science objectives for 100-day missions exploiting human presence and begin technology R&D to expand science capabilities.

- Refine NASA technologies for cellular biotechnology to expand the range of research enabled by space-based facilities.

- Carry out space-based investigations to resolve fundamental issues in low-gravity fluid flow control (including studies of two-phase flows, interfacial phenomena, complex and non-Newtonian fluid characterization).

- Conduct long-duration research on materials and related topics (such as the low-gravity formation of solid materials, the causes and influence of impurities and defects, and the determination of fundamental thermophysical properties of industrial and commercial interest).

- Gain a firm understanding of the fundamental principles that control combustion processes in a variety of fuels; apply that understanding to improve efficiency and reduce hazards on Earth and in space; and resolve the physics and chemistry of combustion-based materials synthesis.

- Perform space-based atomic physics investigations to probe the quantum nature of matter, and pursue applications (such as ultra-fast computer memories).

- Expand our understanding of the effects of fractional gravity on molecular structures, cells, biological systems, organisms, ecosystems, and evolution.

- Influence biological theory and thought through understanding of gravitational and space biology, ecology, and remote sensing.

- Use the ISS centrifuge facility to understand gravitational mechanisms and potential requirements for artificial gravity on long-duration exploration missions.

- Utilize the fundamental knowledge of space biology gained in the near-term to design and conduct ISS experiments that will advance our understanding of biologic processes in space by an order of magnitude (compared to 1998–99 levels).

- Refine scientific objectives to enable definitive investigations of significant unanswered questions.

- Build on the knowledge gained in the near- and mid-term to discover, manipulate, and control biological responses to long-duration space flight.

- Utilize research laboratories on solar system bodies to achieve a deep understanding of the effects of gravity on the evolution, development, and function of living organisms and their environment.

- Develop advanced bioreactors that can support tissue engineering and other biotechnology advances (e.g., organoid production for transplantation and extracorporeal maintenance strategies).

- Develop and use advanced miniaturized microgravity research instrumentation to enable dramatic scientific and technological advances, including studies on unique spacecraft or at in-situ planetary outposts in areas such as resource utilization/generation, extremophile research, synthetic biomolecular machines, and biosensing/biosentinel systems.

- Develop and utilize advanced physics instrumentation (such as laser tweezers, high-speed complex signal processing, and nanoscale sensors) to scrutinize the molecular processes allowing the generation of complex materials systems from simple building blocks, to study cellular motor processes, and to manipulate DNA.

- Advance technologies using improved understanding of combustion processes to achieve higher efficiency conversion of chemical energy to useful work with minimal generation of pollutants, and for the production of new high-value materials.

Enable Humans to Live and Work Permanently in Space

Advances in technology notwithstanding, the human element continues to be the major factor in the success or failure of most terrestrial enterprises. In many cases, innovative technologies are most effective when used to leverage or enhance the productivity of humans. Moreover, the human element is a quintessential component in the public's continuing interest in and support for the space program.

Human presence will be an essential factor in successfully opening the space frontier and expanding knowledge through research in space. As our activities in space grow, so too must human involvement. In this way, we open the door to an array of benefits, tangible and intangible, for the people of the United States and the world.

It is, therefore, a goal of NASA to enable and establish permanent and productive human presence in space, to advance America's aspirations and opportunities in space through new technologies and new ways of doing business.

Our Strategic Objectives

• Provide safe, affordable, and improved access to space

• Operate the International Space Station to advance science, exploration, engineering, and commerce

• Ensure the health, safety, and performance of humans living and working in space

• Meet sustained space operations needs while reducing costs

It is NASA's goal to enable and establish a permanent and productive human presence in space. Here, an astronaut works on construction of the International Space Station.

12

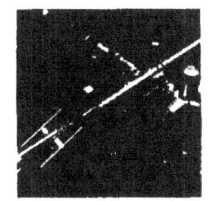

Time-Phased Objectives

- Complete transition of Space Shuttle operations to the Space Flight Operations Contractor and undertake needed Shuttle upgrades consistent with the objectives of increasing safety by about 50 percent and reducing costs per payload pound by 20 percent (compared to late 1990's levels).

- Complete development of the International Space Station—including international partner contributions—to enable a robust ISS research program to begin.

- Pursue conversion of the International Space Station (ISS) to a user-driven operation by creating a non-governmental organization (NGO) to simplify the process for—and reduce the costs of—access to and working in low-Earth orbit.

- Conduct scientific exploration and engineering research—and enable commercial research—activities on the ISS to enrich the health and quality of life on Earth.

- Partner within NASA and with industry to mature technologies needed for future HEDS space transportation—including vehicles and supporting infrastructure, for both space launch and in-space transportation.

- Partner with industry to validate technology to mitigate risk associated with privatization and/or commercialization of space communications systems and/or operations.

- Work with industry to frame scenarios and R&D plans, and to validate competing technologies in space (e.g., on the ISS) for application in 100-day human expeditions beyond low-Earth orbit, including enabling a commercial services approach to mission communications.

- Identify mechanisms of health risk and potential physiological and psychological problems and develop/test countermeasures for these risks (including the radiation environment); and understand the risks/benefits of space mission analogs.

- Test and validate technologies and systems that can reduce the overall mass of human support systems by a factor of two (compared to 1990's levels).

- Conduct research in analog test beds and on orbit to enhance medical care during intravehicular activities, extravehicular activities, and on return to gravity.

- Strengthen partnerships with the Space Shuttle contractors and industry to identify and undertake Space Shuttle upgrades until a credible replacement launch vehicle is available.

- Work in partnership within NASA and with industry to complete technology needed for future HEDS space transport—and inform industry decisions on affordable ground infrastructure for commercial missions.

- Implement commercialization and/or privatization of NASA communications associated with space launch and Earth orbit operations in order to reduce associated Government costs by a factor of 2:1 (compared to mid-1990's levels), and buy commercial communications services for initial 100-day class human missions.

- Undertake pilot efforts leading to commercialization of ISS operations to reduce costs while improving safety and productivity.

- Initiate Government-commercial partnerships in research, development, and infusion of new technology to extend ISS life beyond 2012, as needed.

- Conduct full-scale scientific, exploration and engineering research—and enable commercial research—activities on the ISS within the context of the NGO.

- Partner with industry to build a first generation, advanced in-space transportation system capable of meeting both commercial and 100-day human mission needs, while reducing costs by 3:1 (versus 1990's systems).

- Understand the effects of long-duration space flight (including radiation); validate countermeasures and technology; and test response to gravity analogs; and refine biomedical knowledge and experience to identify and begin developing countermeasures for long-duration space flight.

- Test and validate technologies and systems that can reduce the overall mass of the human support system by a factor of three (compared to 1990's levels).

- Translate terrestrial diagnostic, drug therapy, and minimally invasive surgery to the microgravity environment.

- Complete research and technology validation (including demonstrations on the ISS) of competing technologies for 100- to 1,000-day human missions.

- Partner with industry to reduce HEDS Enterprise space transportation costs more than fivefold while ensuring safety, by building:

    - A next-generation launch capability able to meet HEDS mission needs.

    - Evolutionary in-space transportation systems capable of meeting commercial and human exploration mission needs.

- Perfect countermeasures for safe, effective, and affordable long-duration human space flight.

- Complete the development of safe, self-sufficient, and self-sustaining systems that can enable humans to live and work in space and on other planets—independent from Earth-provided logistics—for extended periods.

- Develop autonomous clinical medicine capabilities for long-duration missions by exploiting advanced technologies (e.g., artificial intelligence, robotic surgery, and virtual consultations).

- Complete the transition of the ISS to a customer-driven commercial operation and work with industry to identify and implement major upgrades as needed to extend ISS life expectancy and/or expand capability to meet user community needs, while improving safety.

- Buy selected commercial communications services for initial 1000-day human missions.

- Buy services from the ISS commercial operator to validate technologies for evolutionary human missions of more than 1,000 days.

Enable the Commercial Development of Space

Commerce is essential to human society; free market transactions are the foundation of the dramatic progress humankind has made during the past several centuries. Wherever humans go and wherever they live, there too is commerce. Moreover, the free market is an effective mechanism for delivering tangible benefits from space broadly to the American people.

If humanity is to explore and develop space, to better exploit the space environment for profound scientific discoveries, and someday to settle the space frontier, it may be through the continuing expansion of the private sector—of individuals and of industry—into space. As we open the space frontier, we must therefore seek to expand the free market into space.

It is a goal of NASA to enable the commercial development of space.

Our Strategic Objectives

- Improve the accessibility of space to meet the needs of commercial research and development

- Foster commercial endeavors with the International Space Station and other assets

- Develop new capabilities for human space flight and commercial applications through partnerships with the private sector

Commercial product development modules are in the process of being installed and attached to the external structure of the International Space Station.

Time-Phased Objectives

- Pursue establishment of a non-governmental organization (NGO) as a partner with NASA to place leadership for fostering commercial endeavors with the International Space Station (ISS) outside Government and simplify the process for—and reduce the cost of—access to space for commercial R&D or other ventures:

  - Permit 30 percent of the U.S. share of ISS pressurized accommodations to be used for commercially sponsored R&D or other ventures and 20 percent of the U.S. share for unpressurized payload accommodations.

  - Ensure Government processing in less than 6 months and negotiation in less than 3 months, on average, of commercial ISS proposals.

- Formulate and advocate policy, legislative and engineering actions as needed to facilitate privately funded commercial space development, including removing barriers and providing appropriate incentives.

- Create new approaches to collaborative partnerships with the private sector for the development of future HEDS Enterprise capabilities.

- Identify (jointly with industry) the commercial potential of candidate exploration concepts and technologies for 100-day class human missions and establish cooperative R&D projects to reduce Government cost substantially.

- Continue to permit 30 percent of the U.S. share of ISS pressurized accommodations to be used for commercially sponsored R&D or other ventures and 20 percent of the U.S. share for unpressurized payload accommodations, working through the ISS NGO, as appropriate.

- Offer the ISS as a commercial R&D test bed with the goal of a ratio of greater than 3:1 in private to public financing.

- Expand ISS operations to permit regular docking of commercial transportation systems and space platforms, while reducing public subsidies to private enterprises on the Station.

- Refine and implement policy, legislative, and engineering actions to enable expanded low-Earth orbit operations through private development, launch, and operation of self-sustaining, co-orbiting infrastructure, which employs the ISS principally as a service center at a low marginal cost.

- Identify (jointly with industry) the commercial potential of various concepts for 1,000-day human exploration missions and establish cooperative R&D projects to develop the candidate technologies.

- Achieve an increase of 2:1 in the total ISS accommodations available for commercially sponsored R&D and other ventures (compared to the mid-term).

- Complete the transition of the ISS from public to private and fully transform the Government role from sole supplier to customer.

- Transition the ISS to private sector commercial operators as a fee-for-service scientific laboratory, technology test bed, and business venue, with all customers trading in a free market economy.

- Transfer responsibility for systems in Earth orbit to the private sector and dedicate NASA R&D to the risks of expanding the frontier.

In the 21st century, NASA's International Space Station will provide many commercial opportunities for business people worldwide.

Share the Experience and Benefits of Discovery

Americans—of all backgrounds—should have the opportunity to share in the experience and benefits of space exploration and development. During the past four decades, ambitious human space flight missions have inspired generations of young people to undertake careers in science, mathematics, and engineering—benefiting both themselves and society. The space program can enrich society by directly enhancing the quality of education.

Terrestrial applications of technologies developed for space have saved many lives, made possible medical breakthroughs, created countless jobs, and yielded diverse other tangible benefits for Americans. The further commercial development of space will yield still more jobs, technologies, and capabilities to benefit people the world over in their everyday lives.

A goal of NASA is therefore to share the experience, the excitement of discovery, and the benefits of human space flight with all.

Our Strategic Objectives

- Engage and involve the public in the excitement and the benefits of—and in setting the goals for—the exploration and development of space

- Provide significantly more value to significantly more people through exploration and space development efforts

- Advance the scientific, technological, and academic achievement of the Nation by sharing our knowledge, capabilities, and assets

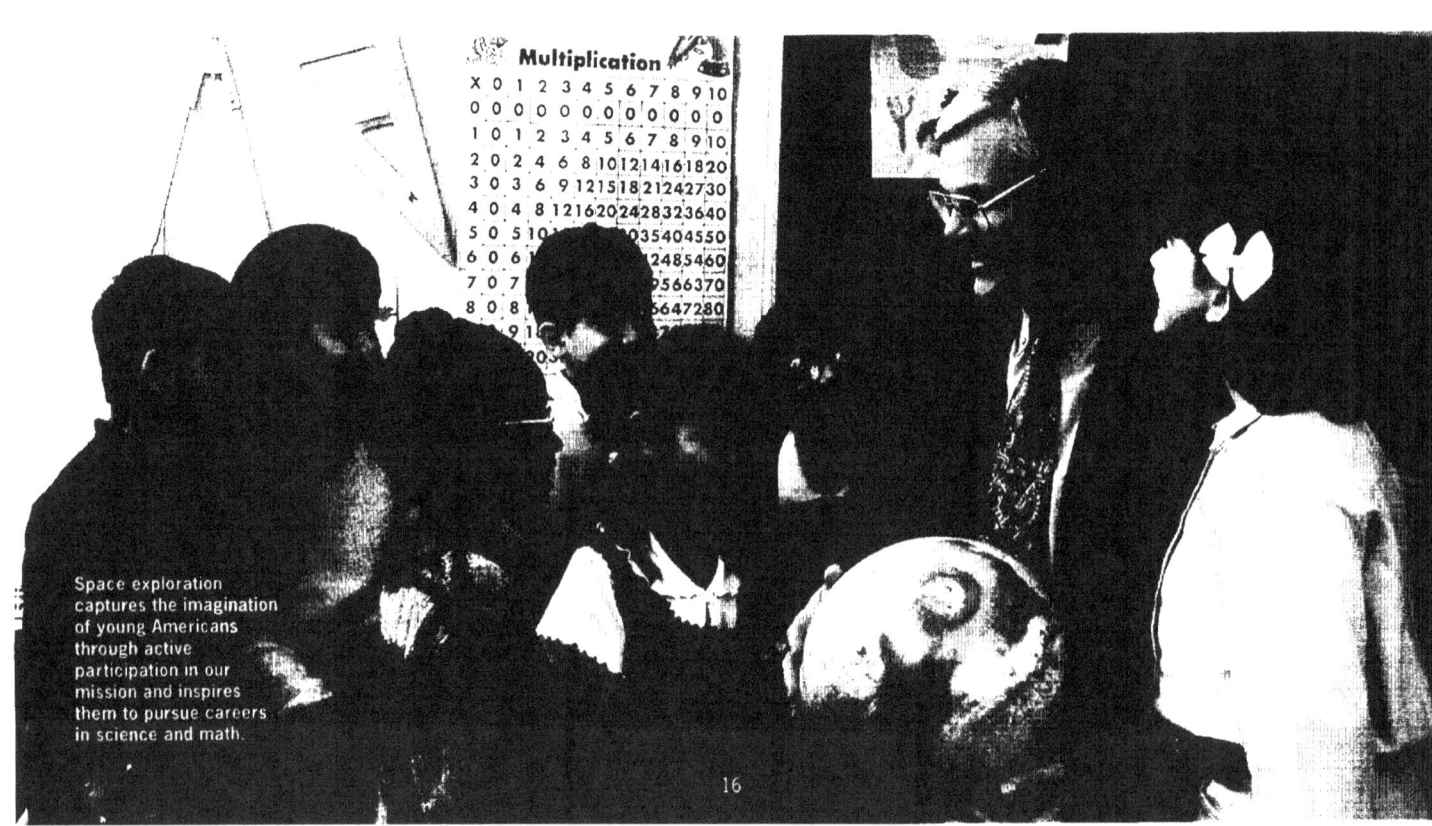

Space exploration captures the imagination of young Americans through active participation in our mission and inspires them to pursue careers in science and math.

Time-Phased Objectives

- Establish a focused customer engagement process to significantly increase our value to significantly more people by enabling the public to guide the formulation of HEDS goals, objectives, programs, and missions, and develop a "Roadmap to the Stars" effort to communicate broadly the new knowledge, breakthrough technologies, and innovative capabilities associated with various prospective HEDS Enterprise activities.

- Expand public access to HEDS mission information (especially the ISS) by working with industry to create media projects and public engagement initiatives that allow "first-hand" public participation using telepresence for current missions, and virtual reality or mock-ups for future missions beyond Earth orbit.

- Provide appropriate NASA support for broad industrial development in space, including public space travel, in the context of increasing commercialization of space operations and development.

- Build partnerships with educators at all levels—colleges and universities (including Historically Black Colleges and Universities and similar institutions) and K–12—in order to:

  - Increase the involvement of faculty and students in establishing HEDS goals and objective.

  - Enable "first-hand," interactive participation in both human and robotic exploration.

  - Make exploration a part of new science curricula.

  - Create a pool of capable and diverse graduates.

- Analyze and translate HEDS science results to support curriculum and professional development standards for the K–12 education community.

- Use the HEDS Enterprise's unique mission to encourage K–12 teachers and students to improve science literacy and to help them incorporate science, mathematics, technology, and engineering disciplines into course work. Programs will emphasize teacher preparation and enhancement and the participation of underrepresented groups.

- Prepare the next generation of researchers through programs focused on undergraduate and graduate students in science and engineering programs (particularly underrepresented groups), increasing the opportunity for "hands-on" experience in HEDS-related disciplines.

- Seek opportunities with informal educational institutions (e.g., museums) to use HEDS and related science results and products in order to foster the development of an informed and aware public.

- Work with college and university faculty and students in the conducting of HEDS research and technology for future exploration.

- Continue human exploration customer engagement initiatives that allow "first-hand" participation in HEDS missions beyond Earth orbit—including both robotic "outpost" activities and 100-day class human expeditions being studied or implemented—using telepresence virtual reality or mock-ups, as appropriate.

- Strengthen and expand partnerships with colleges and universities (including Historically Black Colleges and Universities and similar institutions) and K–12 to:

  - Assure continuing involvement in establishing HEDS Enterprise goals and objectives.

  - Broaden "first-hand," interactive participation in exploration, including developing "outposts" and 100-day class human missions.

  - Continue to make exploration a part of new science curricula.

  - Extend the pool of capable and diverse graduates in the 2010+ timeframe.

- Strengthen and expand relationships with colleges and universities in the implementation of HEDS R&T.

- Expand education opportunities with K–12 teachers and students and informal education institutions by exploiting the new knowledge, new technologies, and discoveries from the ISS, collaborative robotic missions, and the first human missions beyond Earth orbit to inspire students to pursue science and technology curricula.

- Expand public engagement through continuing initiatives that allow meaningful participation in ongoing or planned HEDS missions beyond Earth orbit—including both robotic "outpost" activities and 100- to 1,000-day class human expeditions.

- Retarget HEDS Enterprise academic research and exchange programs to focus on ambitious integrated human-robotic exploration missions, involving colleges, universities (including Historically Black Colleges and Universities and similar institutions), and K–12.

- Update HEDS education programs to exploit the most recent discoveries from human and/or robotic exploration missions to inspire students to pursue science and technology curricula.

Implementing Strategies

In addition to the specific strategic goals and objectives described, the HEDS Enterprise must also undertake the following general goals in order to create a foundation for success:

- Engage NASA's customers in setting HEDS goals, objectives, and priorities

- Ensure that safety and health are inherent in all that we undertake

- Focus on research and development, and invest in breakthrough technologies

- Privatize and commercialize operational activities

- Employ open, competitive processes for selecting research projects

- Promote synergy with other Enterprises and cooperation and engagement with organizations and customer communities outside of NASA

- Promote synergy between fundamental research disciplines and mission-oriented research within HEDS and with other Enterprises

- The HEDS Enterprise must forge partnerships and customer engagement alliances across a broad spectrum, including:

  - Academia

  - Industry

    - Aerospace

    - Non-Aerospace (e.g., engineering, information technology, health care, biotechnology, electronics)

  - Other NASA Enterprises:

    - Space Science

    - Earth Science

    - Aero-Space Technology

  - Other Organizations, including:

    - International Space Agencies and Organizations

    - Other U.S. Government agencies

    - Non-profit and non-governmental agencies

Important Areas of Scientific Inquiry

• What is the role that gravity plays in biological processes of both plants and animals?

• What is the structure of biological molecules that play a critical role in health and disease?

• How do human tissues form organized structures?

• How can humans live and function productively in an environment away from the surface of the Earth?

• How do small particles interact to form ordered systems in diverse applications ranging from colloid-based optical switches to earthquakes?

• How can heat transfer systems be made more safe and efficient?

• What is the interplay between turbulence, chemistry, and heat transfer in soot formation during combustion?

• How can we use the low gravity of space to test the laws of physics to limits that are unachievable on Earth?

• Does life exist elsewhere in our solar system? Did it exist in the past?

• What are the resources of the solar system? Where are they? Are they accessible for human use? How have they changed with time?

• And finally, how can we apply advances in our understanding in these areas—both in space and terrestrially?

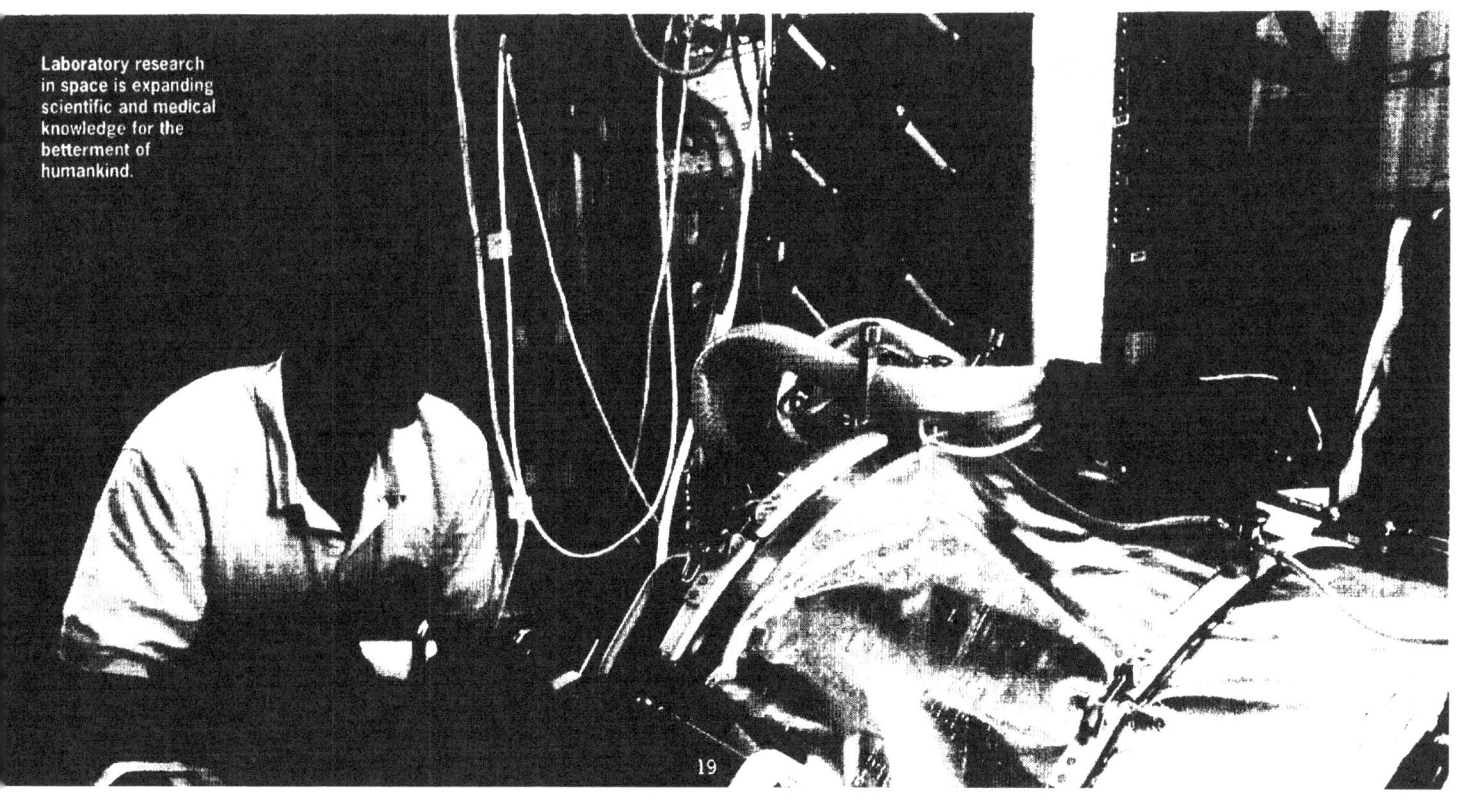

Laboratory research in space is expanding scientific and medical knowledge for the betterment of humankind.

Challenges

The goals and objectives identified in the HEDS Strategic Plan are responsive to sustaining these

- Engage our customers in setting goals

- Ensure and enhance reliability and operational safety in all ground, flight, and on-orbit activities

- Successfully complete ISS development and demonstrate sustained and productive human activities in space

- Reduce costs of human access to—and operations in—space

- Reduce transit times and dependence on Earth

- Improve opportunities for scientific return

- Commercialize routine space operations while increasing commercial participation and opportunities to provide services to Government missions

- Increase commercial investment in the capabilities needed to achieve exploration and scientific research goals

- Reduce costs, time delays, and uncertainties associated with access to—and operations in—space for commercial activities

- Inspire the public through HEDS programs

- Increase broad public access to near-real-time information while providing enhanced and more varied information and knowledge sources

- Enlarge the pool of talented scientists and engineers to undertake the human exploration and development of space—both in Government and the private sector

- Accomplish more effective transition of technologies to new applications

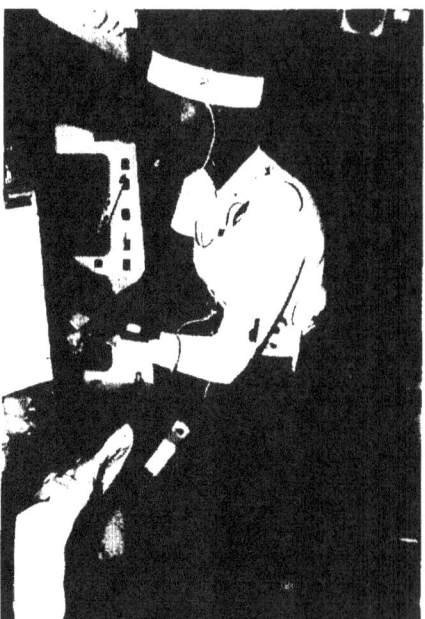

Conclusion

During the coming 25 years, we hope to achieve truly profound goals in space.

In so doing, we can make possible the permanent extension of human presence beyond the bounds of Earth and enable historic improvements in the quality of life and in our understanding of the solar system and the universe.

By implementing this Strategic Plan for the Human Exploration and Development of Space Enterprise, we can answer this challenge.

NASA will:

• Continue to fly Space Shuttle safely during its transition to customer-focused privatized operations—with upgrades to assure continuing improvements in safety, reliability, and cost until a viable replacement vehicle is available.

• Complete International Space Station development—creating a new, unprecedented research laboratory— and transition the Station to totally customer-driven commercial operations.

• Define breakthrough concepts for affordable human exploration missions.

• Conduct collaborative robotic missions that lead the way in collecting data, demonstrating technologies, and setting in place continuing operations at key sites (e.g., "outposts").

• Address fundamental scientific questions in biology, medicine, materials, engineering, and other areas, and facilitate solving the challenges of microgravity and radiation for long-duration human space flight.

• Pursue exploration technology R&D through NASA-industry partnerships, while further partnering with industry to enable human exploration and commercial space goals to support one another through the development of important new capabilities that can meet both needs.

• Transform NASA's relationship with students, faculty, industry, and the public by engaging them broadly in setting goals for the human exploration and development of space.

• Make possible dramatic increases in discovery, scientific knowledge, and human accomplishment through U.S.-led international partnerships to develop the capabilities needed to implement:

  - 100-day class human exploration missions (i.e., within the near-Earth region) during the mid-term, and

  - 1,000-day class human exploration missions (i.e., within ~ 2 astronomical units of the Sun) during the far-term.

Engagement

The HEDS Enterprise is determined
to better engage its customers in the
formulation of strategic goals, objectives,
and future program directions.

To begin that process, NASA invites
comments from any interested internal
or external reader of this plan.

Input will be reviewed and considered for
inclusion in a future update of the plan.

heds-strategic-feedback@hq.nasa.gov

Editorial Group

Darrell Branscome (Lead)
NASA Headquarters
Deputy Associate
Administrator for
Enterprise Development
Office of Space Flight
300 E Street SW
Washington, DC 20546

Marc Allen
NASA Headquarters
Office of Space Science
300 E Street SW
Washington, DC 20546

William Bihner
NASA Headquarters
Office of Space Science
300 E Street SW
Washington, DC 20546

Douglas Cooke
Johnson Space Center
2101 NASA Road One
Houston, TX 77058

Mark Craig
Stennis Space Center
Deputy Director of
Stennis Space Center
Stennis Space Center,
MS 39529

Matthew Crouch
NASA Headquarters
Office of Policy and Plans
Strategic Planning
300 E Street SW
Washington, DC 20546

Roger Crouch
Senior Scientist
Office of Life and
Microgravity Sciences
and Applications
300 E Street SW
Washington, DC 20546

Chris Flaherty
NASA Headquarters
Office of Life and
Microgravity Sciences
and Applications
300 E Street SW
Washington, DC 20546

Norman Haynes
Jet Propulsion Laboratory
Office of the Director
4800 Oak Grove Drive
Pasadena, CA 91109

Steven Horowitz
Goddard Space Flight
Center
Office of the Associate
Director
Greenbelt Road
Greenbelt, MD 20771

Pedro Jimenez
NASA Headquarters
Office of Space Flight
300 E Street SW
Washington, DC 20546

John Mankins (Lead Author)
NASA Headquarters
Office of Space Flight
300 E Street SW
Washington, DC 20546

Jo Ann Morgan
Kennedy Space Center
Advanced Development
and Shuttle Upgrades
Kennedy Space Center,
FL 32899

David Morrison
Ames Research Center
Office of Director of Space
Moffett Field, CA 94035

Arthur Murphy
Jet Propulsion Laboratory
Technology and
Applications Program
Directorate
4800 Oak Grove Drive
Pasadena, CA 91109

John J. Nieberding
Glenn Research Center
Senior Advisor for
Advanced Concepts
21000 Brookpark Road
Cleveland, OH 44135

Axel Roth
Marshall Space Flight
Center
Director of Flight Projects
Marshall Space Flight
Center, AL 35812

Donna Shortz
NASA Headquarters
Office of Space Flight
300 E Street SW
Washington, DC 20546

William Smith
Langley Research Center
Office of the Deputy
Director Space Access and
Exploration Program Office
Hampton, VA 23681

John P. Sumrall
NASA Headquarters
Office of Aero-Space
Technology
300 E Street SW
Washington, DC 20546

Senior Management Concurrence

Joseph H. Rothenberg
Associate Administrator
Space Flight

Arnauld E. Nicogossian, M.D.
Associate Administrator
Life and Microgravity
Sciences and Applications

Gen. Spence M. Armstrong
Associate Administrator
Aero-Space Technology

Edward J. Weiler, Ph.D.
Associate Administrator
Space Science

Lori Garver
Associate Administrator
Policy and Plans

Henry McDonald, Ph.D.
Director
Ames Research Center

Donald J. Campbell
Director
Glenn Research Center

Alphonso V. Diaz
Director
Goddard Space Flight
Center

Edward Stone, Ph.D.
Director
Jet Propulsion Laboratory

George W.S. Abbey
Director
Lyndon B. Johnson Space
Center

Gen. Roy D. Bridges
Director
John F. Kennedy Space
Center

Jeremiah F. Creedon, Ph.D.
Director
Langley Research Center

Arthur G. Stephenson
Director
George C. Marshall Space
Flight Center

Roy S. Estess
Director
John C. Stennis Space Center

Please direct all comments and
suggestions to a special e-mail address:

heds-strategic-feedback@hq.nasa.gov

www.ingramcontent.com/pod-product-compliance
Lightning Source LLC
Chambersburg PA
CBHW081822170526
45167CB00008B/3500